The Story of a T

Charles Meymott Tidy

Alpha Editions

This edition published in 2024

ISBN : 9789362927965

Design and Setting By
Alpha Editions
www.alphaedis.com
Email - info@alphaedis.com

Contents

PREFACE.

These lectures were delivered with the assistance merely of a few notes, the author in preparing them for the press adhering as nearly as possible to the shorthand writer's manuscript. They must be read as intentionally untechnical holiday lectures intended for juveniles. But as the print cannot convey the experiments or the demonstrations, the reader is begged to make the necessary allowance.

The author desires to take this opportunity of expressing his thanks to Messrs. Bryant and May; to Messrs. Woodhouse and Rawson, electrical engineers; to Mr. Woolf, the lead-pencil manufacturer; and to Mr. Gardiner, for numerous specimens with which the lectures were illustrated.

LECTURE I.

MY YOUNG FRIENDS,—Some months ago the Directors of this Institution honoured me with a request that I should deliver a course of Christmas Juvenile Lectures. I must admit I did my best to shirk the task, feeling that the duty would be better intrusted to one who had fewer demands upon his time. It was under the genial influence of a bright summer's afternoon, when one thought Christmas-tide such a long way off that it might never come, that I consented to undertake this course of lectures. No sooner had I done so than I was pressed to name a subject. Now it is a very difficult thing to choose a subject, and especially a subject for a course of juvenile lectures; and I will take you thus much into my confidence by telling you that I selected the subject upon which I am to speak to you, long before I had a notion what I could make of it, or indeed whether I could make anything at all of it. I mention these details to ask you and our elders who honour us—you and me—with their company at these lectures, for some little indulgence, if at times the story I have to tell proves somewhat commonplace, something you may have heard before, a tale oft told. My sole desire is that these lectures should be true *juvenile* lectures.

Well, you all know what this is? [*Holding up a box of matches.*] It is a box of matches. And you know, moreover, what it is used for, and how to use it. I will take out one of the matches, rub it on the box, and "strike a light." You say that experiment is commonplace enough. Be it so. At any rate, I want you to recollect that phrase—"strike a light." It will occur again in our course of lectures. But, you must know, there was a time when people wanted fire, but had no matches wherewith to procure it. How did they obtain fire? The necessity for, and therefore the art of producing, fire is, I should suppose, as old as the world itself. Although it may be true that our very earliest ancestors relied for necessary food chiefly on an uncooked vegetable diet, nevertheless it is certain that very early in the history of the world people discovered that cooked meat (the venison that our souls love) was a thing not altogether to be despised. Certainly by the time of Tubal Cain, an early worker in metals, not only the methods of producing fire, but also the uses to which fire could be applied, must have been well understood. Imagine the astonishment of our ancestors when they first saw fire! Possibly, the first sight of this wonderful "element" vouchsafed to mortals was a burning mountain, or something of that kind. One is scarcely astonished that there should have been in those early times a number of people who were professed fire-worshippers. No wonder, I say, that fire should have been regarded with intense reverence. It constituted an essential part of early sacrificial worship. Some of my young friends, too, may remember how in ancient Rome there was a special order (called the order of the Vestal Virgins), whose duty it was to preserve the sacred fire, which if once extinguished, it was thought would bring ruin and destruction upon their city.

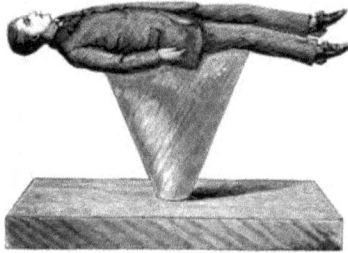

Fig. 1.

How did our ancestors, think you, obtain fire in those early times? I suggested a burning mountain as a source of fire. You remember, too, perhaps reading about Prometheus, who stole fire from heaven, bringing it to earth in a copper rod, which combined act of theft and scientific experiment made the gods very angry, because they were afraid mortals might learn as many wonderful things as they knew themselves. History seems to show that the energetic rubbing together of dry sticks was one of the earliest methods adopted by our ancestors for producing fire. I find, for instance, described and pictured by an early author some such plan as the following:—A thick piece of wood was placed upon the ground. Into a hole bored in this piece of wood a cone of wood was fitted. By placing a boy or man on the top of the cone, and whirling him round, sufficient friction resulted where the two pieces of wood rubbed one against the other to produce fire. Our artist has modernized the picture to give you an idea of the operation (Fig. 1). Now instead of repeating that experiment exactly, I will try to obtain fire by the friction of wood with wood. I take this piece of boxwood, and having cut it to a point, rub it briskly on another piece of wood (Fig. 2). If I employ sufficient energy, I have no doubt I may make it hot enough to fire tinder. Yes! I have done so, as you see. (I will at once apologize for the smoke. Unfortunately we cannot generally have fire without smoke.) Every boy knows that experiment in another form. A boy takes a brass button, and after giving it a good rub on his desk, applies it to the cheek of some inoffensive boy at his side, much to the astonishment of his quiet neighbour. Well, I am going to see whether I can produce fire with a brass button. I have mounted my button, as you see, for certain reasons on a cork, and I will endeavour by rubbing the button on a piece of pinewood to make it sufficiently hot to fire tinder. Already I have done so.

Fig. 2.

Talking about friction as a means of producing heat, I should like to mention that at the last Paris Exhibition I saw water made to boil, and coffee prepared from it, by the heat resulting from the friction of two copper plates within the liquid.

That then is the earliest history I can give you of the production of fire, and at once from that history I come to the reign of the tinder-box. The tinder-box constitutes one of the very earliest methods, no doubt, of obtaining fire. I have searched for some history of the tinder-box, and all I can say for certain is that it was in use long before the age of printing. I have here several rare old tinder boxes. I intend showing you in the course of these lectures every detail of their construction and use. I have no doubt this very old tinder-box that you see here (Fig. 3 A) was once upon a time kept on the mantel-piece of the kitchen well polished and bright, and I do not doubt but that it has lit hundreds and thousands of fires, and, what is more, has very often been spoken to very disrespectfully when the servant wanted to light the fire, and her master was waiting for his breakfast. I will project a picture of it on the screen, so that you may all see it. There it is. It is a beautiful piece of apparatus. There is the tinder, the steel (Fig. 3 *b*), the flint (*c*), and the matches (*d*) complete.

Fig. 3.

It was with this instrument, long before the invention of matches, that our grandfathers obtained light. I want to show you how the trick was managed. First of all it was necessary to have good tinder. To obtain this, they took a piece of linen and simply charred or burnt it, as you see I am doing now (Fig. 4). (Cambric, I am told, makes the best tinder for match-lighting, and the ladies, in the kindness of their hearts, formerly made a point of saving their old cambric handkerchiefs for this purpose.) The servants prepared the tinder over-night, for reasons I shall explain to you directly. Having made the tinder, they shut it down in the box with the lid (Fig. 3 A) to prevent contact with air. You see I have the tinder now safely secured in my tinder-box. Here is a piece of common flint, and here is the steel. Here too are the matches, and I am fortunate in having some of the old matches made many years ago, prepared as you see with a little sulphur upon their tips. Well, having got all these etceteras, box, tinder, flint and steel, we set to work in this way:—Taking the steel in one hand, and the flint in the other, I must give the steel a blow, or rather a succession of blows with the flint (Fig. 3 B). Notice what beautiful sparks I obtain! I want one of these sparks, if I can persuade it to do so, to fall on my tinder. There! it has done so, and my tinder has caught fire. I blow my fired tinder a little to make it burn better, and now I apply a sulphur match to the red-hot tinder. See, I have succeeded in getting my match in flame. I will now set light to one of these old-fashioned candles—a rushlight—with which our ancestors were satisfied before the days of gas and electric lighting. This was their light, and this was the way they lighted it. No wonder (perhaps you say) that they went to bed early.

Fig. 4.

I should like to draw your attention to one other form of tinder-box, because I do not suppose you have ever seen these kind of things before. I have here two specimens of the pistol form of tinder-box (Fig. 5). Here is the flint, the tinder being contained in this little box. It is the same sort of tinder as we made just now. The tinder was fired with flint and steel in the same way as the old-fashioned flint pistols fired the gunpowder. And you see this pistol tinder-box is so constructed as to serve as a candlestick as well as a tinder-box. I have fired, as you perceive, my charred linen with this curious tinder-box, and thus I get my sulphur match alight once more!

Fig. 5.

It was in the year 1669 that Brandt, an alchemist and a merchant—a very distinguished scientific man—discovered the remarkable substance I have here, which we call phosphorus. Brandt was an alchemist. I do not know whether you know what an alchemist is. An alchemist was an old-fashioned chemist. These alchemists had three prominent ideas before them. The first thing they sought for was to discover a something—a powder they thought it ought to be—that would change the commoner or baser metals (such as iron) into gold. The second idea was to discover "a universal solvent," that is, a liquid which would dissolve everything, and they hoped out of this liquid to be able to crystallize gems. And then, having obtained gold and gems, the third thing they desired was "a vital elixir" to prolong their lives indefinitely to enjoy the gold and gems they had manufactured. These were the modest aims of alchemy. Well now—although you may say such notions sound very foolish—let me tell you that great practical discoveries had their origin in the very out-of-the-way researches of the alchemists.

Depend upon this, that an object of lofty pursuit, though that object be one of practically impossible attainment, is not unworthy the ambition of the scientific man. Though we cannot scale the summit of the volcanic cone, we may notwithstanding reach a point where we can examine the lava its fires have melted. We may do a great deal even in our attempt to grasp the impossible. It was so with Brandt. He was searching for a something that would change the baser metals into gold, and, in the search, he discovered phosphorus. The chief thing that struck Brandt about phosphorus was its property of shining in the dark without having previously been exposed to light. A great many substances were known to science even at that time that shone in the dark *after* they had been exposed to light. But it was not until Brandt, in the year 1669, discovered phosphorus that a substance luminous in the dark, without having been previously exposed to light, had been observed. I should like, in passing, to show you how beautifully these phosphorescent powders shine after having been exposed to a powerful light. See how magnificently brilliant they are! These, or something like them, were known before the time of Brandt.

Shortly after phosphorus had been discovered, people came to the conclusion that it might be employed for the purpose of procuring artificial light. But I want you to note, that although phosphorus was discovered in 1669 (and the general properties of phosphorus seem to have been studied and were well understood within five years of its discovery), it was not until the year 1833 that phosphorus matches became a commercial success, so that until the year 1833, our old friend the tinder-box held its ground. I will try and give you as nearly as I can a complete list of the various attempts made with the purpose of procuring fire between the years 1669 and 1833.

The first invention was what were called "phosphoric tapers." From the accounts given (although it is not easy to understand the description), phosphoric tapers seem to have been sulphur matches with a little piece of phosphorus enclosed in glass fixed on the top of the match, the idea being that you had only to break the glass and expose the phosphorus to air for it to catch fire immediately and ignite the sulphur. If this was the notion (although I am not sure), it is not easy to understand how the phosphoric tapers were worked. The second invention for the purpose of utilizing phosphorus for getting fire was by scraping with a match a little phosphorus from a bottle coated with a phosphorus composition, and firing it by friction. The fact is, phosphorus may be easily ignited by slight friction. If I wrap a small piece of phosphorus in paper, as I am doing now, and rub the paper on the table, you see I readily fire my phosphorus.

Fig. 6.

After this, "Homberg's Pyrophorus," consisting of a roasted mixture of alum and flour, was suggested as a means of obtaining fire. Then comes the "Electrophorus," an electrical instrument suggested by Volta, which was thought at the time a grand invention for the purpose of getting light (Fig. 6 A). The nuisance about this instrument was that it proved somewhat capricious in its action, and altogether declined to work in damp foggy weather. I do not know whether I shall be successful in lighting a gas-jet with the electrophorus, but I will try. I excite this plate of resin with a cat-skin (Fig. 6 B), then put this brass plate upon the resin plate and touch the brass (Fig. 6 C); then take the brass plate off the resin plate by the insulating handle and draw a spark from it, which I hope will light the gas. There, I have done it! (Fig. 6 D.)

Fig. 7.

Well, next after the electrophorus comes the "fire syringe" (Fig. 7). The necessary heat in this case is produced by the compression of air. You see in this syringe stopped at one end, I have a certain quantity of air. My piston-rod (C) fits very closely into the syringe (B), so that the air cannot escape. If I push the piston down I compress the air particles, for they can't get out;—I make them in fact occupy less bulk. In the act of compressing the air I produce heat, and the heat, as you see, fires my tinder.

It was in or about the year 1807 that "chemical matches" were introduced to the public for the first time. These chemical matches were simply sulphur matches tipped with a mixture of chlorate of potash and sugar. These matches were fired by dipping them in a bottle containing asbestos moistened with sulphuric acid. Here is one of these "chemical matches," and here the bottle of asbestos and sulphuric acid. I dip the match into the bottle and, as you see, it catches fire.

Fig. 8.

In the year 1820, Dobereiner, a very learned man, discovered a method of getting fire by permitting a jet of hydrogen to play upon finely-divided platinum. The platinum, owing to a property it possesses in a high degree (which property however is not special to platinum), has the power of coercing the union of the hydrogen and oxygen. Here is one of Dobereiner's original lamps (Fig. 8). I am going to show you the experiment, however, on a somewhat larger scale than this lamp permits. Here I have a quantity of fine platinum-wire, made up in the form of a rosette. I place this over the coal-gas as it issues from the gas-burner, and, as you see, the platinum begins to glow, until at last it becomes sufficiently hot to fire the gas (Fig. 9).

In the year 1826 what were called "lucifers" were invented, and I show you here some of the original "lucifers." They are simply sulphur matches tipped with a mixture of chlorate of potash and sulphide of antimony, and were ignited by drawing them briskly through a little piece of folded glass-paper.

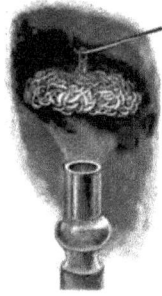

Fig. 9.

In the year 1828, "Prometheans" were invented. I have here two of the original "Prometheans." They consist (as you see) of a small quantity of chlorate of potash and sugar rolled up tightly in a piece of paper. Inside the paper roll is placed a small and sealed glass bubble containing sulphuric acid. When it was wanted to light a "Promethean" you had only to break the bulb of sulphuric acid, the action of which set fire to the mixture of chlorate of potash and sugar, which ignited the paper roll.In the year 1830 "matches" with sulphur tips were introduced as a means of obtaining fire. They were fired, so far as I can make out, by dipping them into a bottle containing a little phosphorus, which then had to be ignited by friction.So far as I know, I have now given you very shortly the history of obtaining fire between the years 1669 and 1830. You see how brisk ingenuity had been during this long period, and yet nothing ousted our old friend the tinder-box. The tinder-box seems, as it were, to speak to us with a feeling of pride and say, "Yes, all you have been talking about were the clever ideas of clever men, but I lived through them all; my flint and my steel were easily procured, my ingredients were not dangerous, and I was fairly certain in my action."In the year 1833 the reign of the tinder-box came to an end. It had had a very long innings—many, many hundred years; but in 1833 its reign was finished. It was in this year the discovery was announced, that bone could be made to yield large quantities of phosphorus at a cheap rate. Originally the price of phosphorus was sufficient to prevent its every-day use. Hanckwitz thus advertises it—"For the information of the curious, he is the only one in London who makes inflammable phosphorus that can be preserved in water. All varieties unadulterated. Sells wholesale and retail. Wholesale, 50s. per oz.; retail, £3 sterling per oz. Every description of good drugs. My portrait will be distributed amongst my customers as a keepsake."

Fig. 10.

Let me give you a brief account of the method of preparing lucifer matches, and to illustrate this part of my story, I am indebted to Messrs. Bryant and May for specimens. Pieces of wood are cut into blocks of the size you see here (Fig. 10 A). These blocks are then cut into little pieces, or splints, of about one-eighth of an inch square (Fig. 10 b). By the bye, abroad they usually make their match splints round by forcing them through a circular plate, pierced with small round holes. I do not know why we in England make our matches square, except for the reason that Englishmen are fond of doing things on the square. The next part of the process is to coat the splints with paraffin or melted sulphur. The necessity for this coating of sulphur or paraffin you will understand by an experiment. If I take some pieces of phosphorus and place them upon a sheet of cartridge paper, and then set fire to the pieces of phosphorus, curiously enough, the ignited phosphorus will not set fire to the paper. I have taken five little pieces of phosphorus (as you see), so as to give the paper every chance of catching fire (Fig. 11). Now that is exactly what would happen if paraffin (or some similarly combustible body) was not placed on the end of the splint; my phosphorus would burn when I rubbed it on the box, but it would not set fire to the match. It is essential, therefore, as you see, in the first instance, to put something on the match that the ignited phosphorus will easily fire, and which will ignite the wood. I will say no more about this now, as I shall have to draw your attention to the subject in another lecture. The end of the splints are generally scorched by contact with a hot plate before they are dipped in the paraffin, after which the phosphorus composition is applied to the match. This composition is simply a mixture of phosphorus, glue, and chlorate of potash. The composition is spread upon a warm plate, and the matches dipped on the plate, so that a small quantity of the phosphorus mixture may adhere to the tip of the match. Every match passes through about seventeen people's hands before it is finished. I told you that in England we generally use chlorate of potash in the preparation of the phosphorus composition, whilst abroad nitrate of potash is usually employed. You know that when we strike a light with an English match a slight snap results, which is due to the chlorate of potash in the match. In the case of nitrate of potash no such snapping noise occurs. Some people are wicked enough to call them "thieves' matches." Just let me show you (in passing) how a mixture of chlorate of potash and sulphur explodes when I strike it.

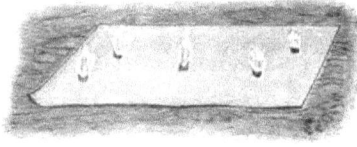

Fig. 11.

Now, then, comes a very remarkable story to which I desire to draw your attention. There were many disadvantages in the use of this yellow phosphorus. First of all, it is a poisonous substance; and what is more, the vapour of the phosphorus was liable to affect the workpeople engaged in the manufacture of lucifer matches with a bad disease of the jaw, and which was practically, I am afraid, incurable. A very great chemist, Schrötter, discovered that phosphorus existed under another form, some of which I have here. This, which is of a red colour, was found to be exactly the same chemical substance as the yellow phosphorus, but possessing in many respects different properties. For instance, you see I keep this yellow phosphorus under water; I don't keep the red phosphorus in water. Amongst other peculiarities it was found that red phosphorus was not a poison, whilst the yellow phosphorus was, as I told you, very poisonous indeed. About two to three grains of yellow phosphorus is sufficient to poison an adult. I have known several cases of children poisoned by sucking the ends of phosphorus matches. So you see it was not unimportant for the workpeople, as well as for the public generally, that something should be discovered equally effective to take the place of this poisonous yellow phosphorus.

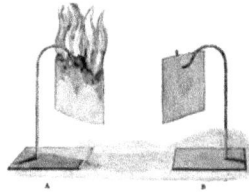

Fig. 12.

I should like to show you what very different properties these two kinds of phosphorus possess. For instance, if I take a small piece of the yellow phosphorus and pour upon it a little of this liquid—bi-sulphide of carbon—and in another bottle treat the red phosphorus in a similar way, we shall find the yellow phosphorus is soluble in the liquid, whilst the red is not. I will pour these solutions on blotting-paper, when you will find that the solution of the yellow phosphorus will before long catch fire spontaneously (Fig. 12 A), whilst the solution (although it is not a solution, for the red phosphorus is not soluble in the bi-sulphide of carbon) of the red phosphorus will not fire (Fig. 12 B). Again, if I add a little iodine to the yellow phosphorus, you see it immediately catches fire (Fig. 13 a); but the same result does not follow with the red phosphorus (Fig. 13 b). I will show you an experiment, however, to prove, notwithstanding these different properties, that this red and yellow material are the same elementary body. I will take a little piece of the yellow phosphorus, and after igniting it introduce it into a jar containing oxygen, and I will make a similar experiment with the red phosphorus. You will notice that the red phosphorus does not catch fire quite so readily as the yellow. However, exactly the same result takes place when they burn—you get the same white smoke with each, and they combust equally brilliantly. The red and yellow varieties are the same body—that is what I want to show you—with different properties.

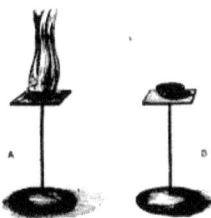

Fig. 13.

Then comes the next improvement in the manufacture of matches, which is putting the phosphorus on the box and not on the match. This is why the use of red phosphorus, was introduced into this country by Messrs. Bryant and May. I have no doubt that many a good drawing-room paper has been spared by the use of matches that light only on the box.I cannot help thinking that the old tinder-box, which I have placed on the table in a prominent position before you to-night, feels a certain pleasure in listening to our story. Envious perhaps a little of its successor, it nevertheless fully recognizes that its own reign had been a thousand times longer than that of the lucifer match. If we could only hear that tinder-box talk, I think we should find it saying something of this kind to the lucifer match—"I gave way to you, because my time was over; but mind, your turn will come next, and you will then have to give way to something else, as once upon a time I had to give way to you." And that is the end of the first chapter of my story of a tinder-box.

LECTURE II.

We were engaged in our last lecture in considering the various methods that have been adopted from early times for obtaining fire, and we left off at the invention of the lucifer match. I ventured to hint at the conclusion of my last lecture, that the tinder-box had something to say to the lucifer match, by way of suggestion, that just as the lucifer match had ousted it, so it was not impossible that something some day might oust the lucifer match. Electricians have unlimited confidence (I can assure you) in the unlimited applications of electricity:—they believe in their science. Now one of the effects of electricity is to cause a considerable rise of temperature in certain substances through which the electrical current is passed. Here is a piece of platinum wire, for example, and if I pass an electrical current through it, you see how the wire glows (Fig. 14). If we were to pass more current through it, which I can easily do, we should be able to make the platinum wire white hot, in which condition it would give out a considerable amount of light. There is the secret of those beautiful incandescent glow lamps that you so often see now-a-days (Fig. 15). Instead of a platinum wire, a fine thread of carbon is brought to a very high temperature by the passage through it of the electrical current, in which condition it gives out light. All that you have to do to light up is to connect your lamp with the battery. The reign of the match, as you see, so far as incandescent electric lamps are concerned, is a thing of the past. We need no match to fire it. Here are various forms of these beautiful little lamps. This is, as you see, a little rosette for the coat. Notice how I can turn the minute incandescent lamp, placed in the centre of the rose, off or on at my pleasure. If I disconnect it with the battery, which is in my pocket, the lamp goes out; if I connect it with my battery the lamp shines brilliantly. This all comes by "switching it on" or "switching it off," as we commonly express the act of connecting or disconnecting the lamp with the source of electricity.

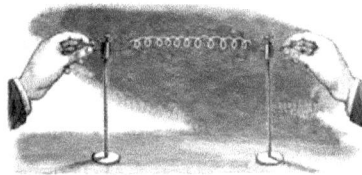

Fig. 14.

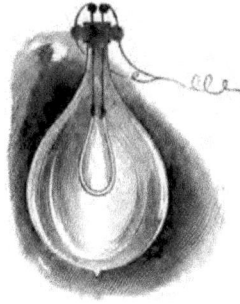

Fig. 15.

Fig. 16.

Here is another apparatus to which I desire to call your attention. If I take a battery such as I have here—a small galvanic battery of some ten cells—you will see a very little spark when I make and break contact of the two poles. This is what is called an electrical torch, in which I utilize this small spark as a gas-lighter (Fig. 16). This instrument contains at its lower part a source of electricity, and if I connect the two wires that run through this long tube with the apparatus which generates the current, which I do by pressing on this button, you see a little spark is at once produced which readily sets fire to my gas-lamp. We have in this electrical torch a substitute—partial substitute, I ought to say—for the lucifer match. I think you will admit that it was with some show of reason I suggested that after all it is possible the lucifer match may not have quite so long an innings as the tinder-box. But there is another curious thing to note in these days of great scientific progress, viz. that there are signs of the old tinder-box coming to the front again. Men, I have often noticed, find it a very difficult thing to light their pipes with a match on the top of an omnibus on a windy day, and inventors are always trying to find out something that will enable them to do so without the trouble and difficulty of striking a match, and keeping the flame a-going long enough to light their cigars. And so we have various forms of pipe-lighting apparatus, of which here is one—which is nothing more than a tinder-box with its flint and steel (Fig. 17). You set to work somewhat in this way: placing the tinder (*a*) on the flint (*b*), you strike the flint with the steel (*c*), and—there, I have done it!—my tinder is fired by the spark. So you see there are signs, not only of the lucifer match being ousted by the applications of electricity, but of the old tinder-box coming amongst us once again in a new form.

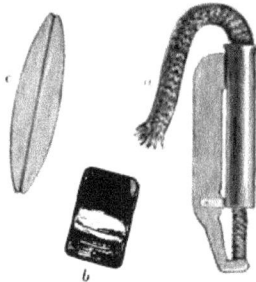

Fig. 17.

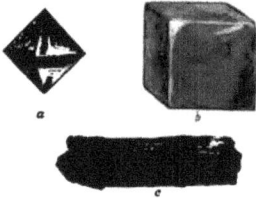

Fig. 18.

I am now going to ask you to travel with me step by step through the operation of getting fire out of the tinder-box. The first thing I have to do is to prepare my tinder, and I told you, if you remember, that the way we made tinder was by charring pieces of linen (see Fig. 4). I told you last time what a dear old friend told me, who from practical experience is far more familiar with tinder-boxes and their working than I am, that no material was better for making tinder than an old cambric handkerchief. However, as I have no cambric handkerchief to operate upon, I must use a piece of common linen rag. I want you to see precisely what takes place. I set fire to my linen (which, by the bye, I have taken care to wash carefully so that there should be no dirt nor starch left in it), and while it is burning shut it down in my tinder-box. That is my tinder. Let us now call this charred linen by its proper name—my tinder is carbon in a state of somewhat fine subdivision. Carbon is an elementary body. An element—I do not say this is a very good definition, but it is sufficiently good for my purpose—an element is a thing from which nothing can be obtained but the element itself. Iron is an element. You cannot get anything out of iron but iron; you cannot decompose iron. Carbon is an element; you can get nothing out of carbon but carbon. You can combine it with other things, but if you have only carbon you can get nothing out of the carbon but carbon. But this carbon is found to exist in very different states or conditions. For instance, it is found in the form of the diamond. (Fig. 18 *a*). Diamonds consist of nothing more nor less than this simple elementary body—carbon. It is a very different form of carbon, no doubt you think, to tinder. Just let me tell you, to use a very hard word, that we call the diamond an "allotropic" form of carbon. Allotropic means an element with another *form* to it—the diamond is simply an allotropic form of carbon. Now the diamond is a very hard substance indeed. You know perfectly well that when the glass-cutter wants to cut glass he employs a diamond for the purpose, and the reason why glass can be cut with a diamond is because the diamond is harder than the glass. I dare say you have often seen the names of people scratched on the windows of railway-carriages, with the object I suppose that it may be known to all future occupants of these carriages that persons of a certain name wore diamond rings. Well, in addition to the diamond there is another form of carbon, which is called black-lead. Black-lead—or, as we term it, graphite—of which I have several specimens here—is simply carbon—an allotrope of carbon—the same elementary substance, notwithstanding, as the diamond. This black-lead (understand black-lead, as it is called, contains no metallic lead) is used largely for making lead-pencils. The manufacture of lead-pencils, by the bye, is a very interesting subject. Formerly they cut little pieces of black-lead out of lumps of the natural black-lead such as you see there; but now-a-days they powder the black-lead, and then compress the very fine powder into a block. There is a block of graphite or black lead, for instance, prepared by simple pressure (Fig. 18 *b*). The great pressure to which the powder is subjected brings these fine particles very close together, when they cohere, and form a substantial block. I will show you an experiment to illustrate what I mean. Here are two pieces of common metallic lead. No ordinary pressure would make these two pieces stick together; but if I push them together very energetically—boys would call it giving them "a shove" together—that is to say, employing considerable pressure to bring them into close contact—I have no doubt that I can make these two pieces of lead stick together—in other words, make them cohere. To cohere is not to adhere. Cohesion is the union of similar particles—like to like; adhesion is the union of dissimilar particles. Now that is exactly what is done in the preparation of the black-lead for lead-pencils. The black-lead powder is submitted to great pressure, and then all these fine particles cohere into one solid lump. The pencil maker now cuts these blocks with a saw into very thin pieces (Fig. 19 *b*). The next thing is to prepare the wood to receive the black-lead strips. To do this they take a piece of flat cedar wood and cut a number of grooves in it, placing one of these little strips of black-lead into each of the grooves (Fig. 19 *a*, which represents one of the grooves). Then having glued on the cover (Fig. 19 *c*), they cut it into strips, and plane each little strip into a round lead-pencil (Fig. 19 *d*). But what you have there as black-lead in the pencil (for this is what I more particularly wish you to remember) is simply carbon, being just the same chemical substance as the diamond. To a chemist diamond and black-lead have the same composition, being indeed the same substance. As to their money value, of course there is some difference; still, so far as chemical composition is concerned, diamonds and black-lead are both absolutely true varieties of the element carbon.

Fig. 19.

Well now, I come to another form of carbon, called charcoal (Fig. 18 *c*). You all know what charcoal is. There is a lump of wood charcoal. It is, as you see, very soft,—so soft indeed is it that one can cut it easily with a knife. Graphite is not porous, but this charcoal is very porous. But mind, whether it be diamond, or black-lead, or this porous charcoal, each and all have the same chemical composition; they are what we call the elementary undecomposable substance carbon. The tinder I made a little while ago (Fig. 4), and which I have securely shut down in my tinder-box, is carbon. It is not a diamond. It is not black-lead, but all the same it is *carbon*—that form of porous carbon which we generally call charcoal. Now I hope you understand the meaning of that learned word *allotropic*. Diamond, black-lead, and tinder are allotropic forms of carbon, just as I explained to you in my last lecture, that the elementary body phosphorus was also known to exist in two forms, the red and the yellow variety, each having very different properties.

Fig. 20.

Now it has been noticed when substances are in a very finely-divided state that they often possess greater chemical activity than they have in lump. Let me try and illustrate what I mean. Here I have a metal called antimony, which is easily acted upon by chlorine. I will place this lump of antimony in a jar of chlorine, and so far as you can see very little action takes place between the metal and the chlorine. There is an action taking place, but it is rather slow (Fig. 20 A). Now I will introduce into the chlorine some of the same metal which I have finely powdered. See! it catches fire immediately (Fig. 20 B). What I want you to understand is, that although I have in both these cases precisely the same chlorine and the same metal, nevertheless, that whilst the action of the chlorine on the *lump* of antimony was not very apparent, in the case of the *powdered* antimony the action was very energetic. Again, there is a lump of lead (Fig. 21 *a*). You would be very much astonished if the lead pipe that conveys the water through your houses caught fire spontaneously; but let me tell you that, if your lead water-pipes were reduced to a sufficiently fine powder, they would catch fire when exposed to the air. I have some finely-powdered lead in this tube (Fig. 21 *b*), which you will notice catches fire directly it is exposed to the atmosphere (Fig. 21 *c*). There it is! Only powder the lead sufficiently fine,— that is to say, bring it into a state of minute subdivision,—and it fires by contact with the oxygen of the air. And now apply this. We have in our diamond the element carbon, but diamond-carbon is a hard substance, and not in a finely-divided state. We have in this tinder the same substance as the diamond, but tinder-carbon is finely divided, and it is because it is in a finely-divided condition that the carbon in our tinder-box catches fire so readily. I hope I have made that part of my subject quite clear to you. I should wish you to note that this very finely-divided carbon has rather an inclination to attract moisture. That is the reason why our tinder is so disposed to get damp, as I told you; and, as damp tinder is very difficult to light, this explains the meaning of those disrespectful words that I suggested our tinder-box had often had addressed to it in the course of its active life of service.

Fig. 21.

But to proceed. What do I want now? I want a spark to fire my tinder. A spark is enough. Do you remember the motto of the Royal Humane Society? Some of my young friends can no doubt translate it, "Lateat scintilla forsan"—perchance a spark may lie hid. If a person rescued from drowning has but a spark of life remaining, try and get the spark to burst into activity. That is what the motto of that excellent society means. How am I to get this spark from the flint and steel to set fire to my tinder? I take the steel in one hand, as you see, and I set to work to strike it as vehemently as I can with the flint which I hold in the other (Fig. 3 A B). Spark follows spark. See how brilliant they are! But I want one spark at least to fall on my tinder. There, I have succeeded, and it has set fire to my tinder. One spark was enough. The spark was obtained by the collision of the steel and flint. The sparks produced by this striking of flint against steel were formerly the only safe light the coal-miner had to light him in his dark dreary work of procuring coal. Here is the flint and steel lamp which originally belonged to Sir Humphry Davy (Fig. 22). The miners could not use candles in coal-mines because that would have been dangerous, and they were driven to employ an apparatus consisting of an iron wheel revolving against a piece of flint for the purpose of getting as much light as the sparks would yield. This instrument has been very kindly lent to me by Professor Dewar. I will project a picture of the apparatus on the screen, so that those at a distance may be better able to see the construction of the instrument.

Fig. 22.

And now follow me carefully. I take the steel and the flint, and striking them together I get sparks. I want you to ask yourselves, Where do the sparks come from? Each spark is due to a minute piece of *iron* being knocked off the steel by the blow of flint with steel. Note the precise character of the spark. Let me sprinkle some iron filings into this large gas flame. You will notice that the sparks of burning iron filings are very similar in appearance to the spark I produce by the collision of my flint and steel.

Fig. 23.

But now I want to carry you somewhat further in our story. It would not do for me simply to knock off a small piece of iron; I want when I knock it off that it should be red-hot. Stay for a moment and think of this—iron particles knocked off—iron particles made red-hot. All mechanical force generates heat.[A] You remember, in my last lecture, I rubbed together some pieces of wood, and they became sufficiently hot to fire phosphorus. On a cold day you rub your hands together to warm them, and the cabmen buffet themselves. It is the same story—mechanical force generating heat! The bather knows perfectly well that a rough sea is warmer than a smooth sea. Why?—because the mechanical dash of the waves has been converted into heat. Let me remind you of the familiar phrase, "striking a light," when I rub the match on the match-box. "Forgive me urging such simple facts by such simple illustrations and such simple experiments. The facts I am endeavouring to bring before you are illustrations of principles that determine the polity of the whole material universe." Friction produces heat. Here is a little toy (cracker) that you may have seen before (Fig. 23). It is scientific in its way. A small quantity of fulminating material is placed between two pieces of card on which a few fragments of sand have been sprinkled (Fig. 23 *a*). The two ends of the paper (*b b*) are pulled asunder. The friction produces heat, the heat fires the fulminate, and off it goes with a crack. And now put this question to yourselves, What produced the friction? Force. What is more, the amount of heat produced is the exact measure of the amount of force used. Heat is a form of force. I must urge you to realize precisely this energy of force. When you sharpen a knife you put oil upon the hone. Why?—When the carpenter saws a piece of wood he greases the saw. Why?—When you travel by train you see the railway-porter running up and down the platform with a box of yellow grease with which he greases the wheels. Why?—The answer to these questions is not far to seek—it is because you want your knife sharpened; it is because you want the saw to cut; it is because you want the train to travel. The carpenter finds sawing hard work, and he does not want the force of the muscles of his arm—his labour, in short—to be converted into heat, and so he greases the saw, knowing that the more completely he prevents friction, the more wood he will cut. It is the force of steam that makes the engine travel. Steam costs money. The engine-driver does not want that steam-force to be converted into heat, because every degree of heat produced means diminished speed of his train; and so the porter greases the wheels. But as you approach the station the train must be stopped. The steam is turned off, and the guard puts on what he calls "the brake." What is the brake? It is a piece of wood so constructed and placed that it can be made to press upon the wheel. Considerable friction results between the wheel and the brake;—heat is produced;—the train gradually comes to a stop. Why? We have now the conversion of that force into heat which a minute ago was being used for the purpose of keeping the train a-going. Given a certain force you can have heat *or* motion; but you cannot have heat *and* motion with the same force in the same amount as if you had them singly. In every-day life, you cannot have your pudding and eat it.

$\underline{A}$ I need scarcely say, that whatever is of any value in the following remarks is derived from that charming book of Professor Tyndall's, *Heat a Mode of Motion.*

Heat then is generated by mechanical force; it is a mode of motion. There was an old theory that heat was material. There was heat, for instance, you were told, in this nail. Suppose I hammer it, it will get hot, and at the same time I shall reduce by hammering the bulk of the iron nail. A pint pot will not hold so much as a quart pot. The nail (you were told) cannot hold so much heat when it occupies a less bulk as it did when it occupied a larger bulk. Therefore if I reduce the bulk of the nail I squeeze out some of the heat. That was the old theory. One single experiment knocked it on the head. It was certain, that in water there is a great deal more entrapped heat—"latent heat" it was called—than there is in ice. If you take two pieces of ice and rub them together, you will find the ice melts—the solid ice changes (that is to say) into liquid water. Where did the heat come from to melt the ice? You could not get the heat *from* the ice, because it was not there, there being admittedly more latent heat in the water than in the ice. The explanation is certain—the heat was the result of the friction. And now let me go to my hammer and nail. I wish to see whether I can make this nail hot by hammering. It is quite cold at the present time. I hope to make the nail hot enough by hammering it to fire that piece of phosphorus (Fig. 24). One or two sharp blows with the hammer suffice, and as you see the thing is done—I have fired the phosphorus. But follow the precise details of the experiment. It was *I* who gave motion to the hammer. *I* brought that hammer on to that nail. Where did the motion go to that I gave the hammer? It went into the nail, and it is that very motion that made the nail hot, and it was that heat which lighted the phosphorus. It was *I* who fired the phosphorus: do not be mistaken, *I* fired the phosphorus. It was my arm that gave motion to the hammer. It was my force that was communicated to the hammer. It was *I* who made the hammer give the motion to the nail. It was *I* myself that fired the phosphorus.

Fig. 24.

I want you then to realize this great fact, that when I hold the steel and strike it with the flint, and get sparks, I first of all knock off a minute fragment of iron by the blow that I impart to it, whilst the force I use in striking the blow actually renders the little piece of detached iron red-hot. What a wonderful thought this is! Look at the sun, the great centre of heat! It looks as if it were a blazing ball of fire in the heavens. Where does the heat of the sun come from? It seems bold to suggest that the heat is produced by the impact of meteorites on the sun. Just as I, for instance, take a hammer and heat the nail by the dash of the hammer on it, so the dash of these meteorites on the sun are supposed to produce the heat so essential to our life and comfort.

Fig. 25.

Let us take another step forward in the story of our tinder-box. Having produced a red-hot spark and set fire to my tinder, I want you to see what I do next. I set to work to blow upon my lighted tinder. You remember, by the bye, that Latin motto of our school-books—*alĕre flammam*, nourish the flame. When I blow on the tinder my object is to nourish the flame. Here is a pair of common kitchen bellows (Fig. 25); when the fire is low the cook blows the fire to make it burn up. What is the object of this blowing operation? It is to supply a larger quantity of atmospheric oxygen to the almost lifeless fire than it would otherwise obtain. Oxygen is the spark's nourishment and life, and the more it gets the better it thrives. Oxygen is an extremely active agent in nourishing flame. If, for instance, I take a little piece of carbon and merely set fire to one small corner of it, and then introduce it into this jar of oxygen, see how brilliantly it burns; you notice how rapidly the carbon is becoming consumed (Fig. 26). In the tinder-box I blow on the tinder to supply a larger amount of oxygen to my spark. A thing to burn under ordinary conditions must have oxygen, and the more oxygen it gets the better it burns. It does not follow that the supply of oxygen to a burning body must necessarily come directly from the air. Here, for instance, I have a squib. I will fire it and put it under water (Fig. 27). You see it goes on burning whether it is in the water or out of it, because one of the materials of which the squib is composed supplies the oxygen. The oxygen is actually locked up inside the squib. When then I blow upon my tinder, my object is to supply more oxygen to it than it would get under ordinary conditions. And, as you see, the more I blow, within certain limits, the more the spark spreads, until now the whole of my tinder has become red-hot. But my time is gone, and we must leave the rest of our story for the next lecture.

Fig. 26.

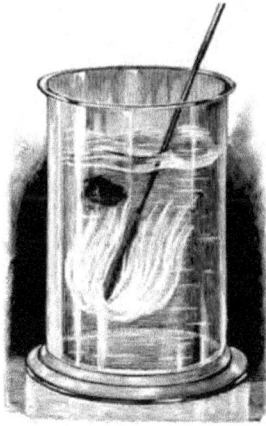

Fig. 27.

LECTURE III.

Recall for a few minutes the facts I brought before you in my last lecture. The first point we discussed was the preparation of the tinder. I explained to you that tinder was nothing more than carbon in a finely-divided state. The second point was, that I had to strike the steel with the flint in such manner that a minute particle of the iron should be detached; the force used in knocking it off being sufficient to make the small particle of iron red-hot. This spark falling upon the tinder set fire to it. The next stage of the operation was to blow upon the tinder, in order, as I said, to nourish the flame; in other words, to promote combustion by an increased supply of oxygen, just as we use an ordinary pair of bellows for the purpose of fanning a fire which has nearly gone out into a blaze.

And now comes the next point in my story of a tinder-box. Having ignited the tinder I want to set fire to the match. Now I have here some of the old tinder-box matches, and you will see that they are simply wooden splints with a little sulphur at the end. Why (you say) use sulphur? For this reason—the wood is not combustible enough to be fired by the red-hot tinder. We put therefore upon the wood a substance which is more combustible than the wood. This sulphur—which most people call brimstone—has been known from very early times. In the middle ages it was regarded as the "principle of fire." It is referred to by Moses and Homer and Pliny. A very distinguished chemist, Geber, describes it as one of "the principles of nature." Having fired my tinder, as you see, and blown upon it, I place my sulphur match in contact with the red-hot tinder. And now I want you to notice that the sulphur match does not catch fire immediately. It wants, in fact, a little time, and as you see a little coaxing. Now I have got it alight. But note, it is the sulphur that at the present moment is burning. The burning sulphur is now beginning to set fire to the wood. The whole match is well alight now! But it was the sulphur that caught fire first, and it was the sulphur that set fire to the wood. A little time was occupied, we said, in making the sulphur catch fire. Ask yourselves this question—Why was it that the sulphur took a little time to catch fire? This was the reason—because before the sulphur could catch fire it was necessary to change the *solid* sulphur (the condition in which it was upon the match end) into *gaseous* sulphur. The solid sulphur could not catch fire. Therefore the heat of my tinder during the interval that I was coaxing the match (as I called it) was being exerted in converting my solid into gaseous sulphur. When the solid sulphur had had sufficient heat applied to it to vapourize it, the sulphur gas immediately caught fire. Now understand, that in order to convert a solid into a liquid, or a liquid into a gas, heat is always a necessity. I must have heat to produce a gas out of a solid or a liquid. I will endeavour to make this clear to you by an experiment. I have here, as you see, a wooden stool, and I am about to pour a little water on this stool. I place a glass beaker on the stool, the liquid water only intervening between the stool and the bottom of the glass. You see the glass is perfectly loose, and easily lifted off the stool notwithstanding the layer of water. I will now pour into the beaker a little of a very volatile liquid—*i. e.* a liquid that is easily converted into a gas—(bisulphide of carbon). I wish somewhat rapidly to effect the change of this liquid bisulphide of carbon into gaseous bisulphide of carbon, and in order to accomplish this object I must have heat. So I take this tube which, as you see, is connected with a pair of bellows, and simply blow on my bisulphide of carbon. This effects the change of the liquid into a gas with great rapidity. Just as I converted my solid sulphur into a gas by the heat of the tinder, so here I am converting this liquid bisulphide of carbon into a gas by the wind from my bellows. But my liquid bisulphide of carbon must get heat somewhere or another in order that the change of the liquid into a gas, that I desire should take place, may be effected; and so, seeing that the water that I have placed between the glass and the stool is the most convenient place from which the liquid can derive the necessary heat, it says, "I will take the heat out of the water." It does so, but in removing the heat from the water it changes the liquid water into solid ice. And see, already the beaker is frozen to the stool, so that I can actually lift up the stool by the beaker (Fig. 28). Understand then why my sulphur match wanted some time and some coaxing before it caught fire, viz. to change this solid sulphur into gaseous sulphur.

Fig. 28.

Fig. 29.

But let us go a step further: why must the solid sulphur be converted into a gas? We want a flame, and whenever we have flame it is absolutely necessary that we should have a gas to burn. You cannot have flame without you have gas. Let me endeavour to illustrate what I mean. I pour into this flask a small quantity of ether, a liquid easily converted into a gas. If I apply a lighted taper to the mouth of the flask, no gas, or practically none, being evolved at the moment, nothing happens. But I will heat the ether so as to convert it into a gas. And now that I have evolved a large quantity of ether gas, when I apply a lighted taper to the mouth of the flask I get a large flame (Fig. 29). There it is! The more gas I evolve (that is, the more actively I apply the heat) the larger is the flame. You see it is a very large flame now. If I take the spirit lamp away, the production of gas grows less and less, until my flame almost dies out; but you see if I again apply my heat and set more gas free, I revive my flame. I want you to grasp this very important fact, upon which I cannot enlarge further now, that given flame, I must have a gas to burn, and therefore heat as a power is needed before I can obtain flame.

Well, you ask me, is that true of all flame? Where is the gas, you say, in that candle flame? Think for a moment of the science involved in lighting a candle. What am I doing when I apply a lighted match to this candle? The first thing I do is to melt the tallow, the melted tallow being drawn up by the capillarity of the wick. The next thing I do is to convert the liquid tallow into a gas. This done, I set fire to the gas. I don't suppose you ever thought so much was involved in lighting a candle. My candle is nothing more than a portable gas-works, similar in principle to the gas-works from which the gas that I am burning here is supplied. Whether it is a lamp, or a gas-burner, or a candle, they are all in a true sense gas-works, and they all pre-suppose the application of heat to some material or another for the purpose of forming a gas which will burn.

Fig. 30.

Before I pass on, I want to refer to the beautiful burner that I have here. It is the burner used by the Whitechapel stall-keepers on a Saturday night (Fig. 30). (Fig. *a* is an enlarged drawing of the burner.) Just let me explain the science of the Whitechapel burner. First of all you will see the man with a funnel filling this top portion with naphtha (*c*). Here is a stop-cock, by turning which he lets a little naphtha run down the tube through a very minute orifice into this small cup at the bottom of the burner (*a*). This cup he heats in a friend's lamp, thereby converting the liquid naphtha, which runs into the cup, into a gas. So soon as the gas is formed—in other words, so soon as the naphtha has been sufficiently heated—the naphtha gas catches fire, the heat being then sufficient to maintain that little cup hot enough to keep up a regular supply of naphtha gas. When the lamp does not burn very well, you will often see the man poking it with a pin. The carbon given off from the naphtha is very disposed to choke up the little hole through which the naphtha runs into the cup, and the costermonger pushes a pin into the little hole to allow the free passage of the naphtha. That, then, is the mechanism of this beautiful lamp of the Whitechapel traders, known as Halliday's lamp.

Now I go to another point: having obtained the gas, I must set fire to it. It is important to note that the temperature required to set fire to different gases varies with the gas. For instance, I will set free in this bottle a small quantity of gas, which fires at a very low temperature. It is the vapour of carbon disulphide. See, I merely place a hot rod into the bottle, and the gas fires at once. If I put a hot rod into this bottle of coal gas, no such effect results, since coal gas requires a very much higher temperature to ignite it than bisulphide of carbon gas. I want almost—not quite—actual flame to fire coal gas. But here is another gas, about which I may have to say something directly, called marsh gas (the gas of coal-mines). This requires a much higher temperature than even coal gas to fire it. I want you to understand that although all gases require heat to fire them, different gases ignite at very different temperatures. Bisulphide of carbon gas, *e. g.*, ignites at a very low temperature, whilst marsh gas requires a very high temperature indeed for its ignition. You will see directly that this is a very important fact. Sulphur gas ignites fortunately at a fairly low temperature, and that is why sulphur is so useful an addition to the wood splint by which to get fire out of the tinder-box.

Fig. 31.

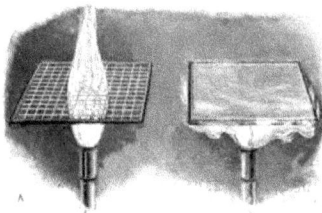

Fig. 32.

And here I wish to make a slight digression in my story. I will show you an experiment preparatory to bringing before you the fact I am anxious now to make clear. I have before me a tube, one half of which is brass and the other half wood. I have covered the tube, as you see, with a tightly-fitting piece of white paper. The whole tube, wood and brass, has been treated in exactly the same manner. Now I will set fire to some spirit in the trough I have here, and expose the entire tube to the action of the flame. Notice this very curious result, viz. that the paper covering the brass portion of the tube does not catch fire, whereas the paper covering the wood is rapidly consumed (Fig. 31). You see the exact line that divides wood from brass by the burning of the paper. Well, why is that? Now all of you know that some things conduct heat (*i. e.* carry away heat) better than other substances. For instance, if you were to put a copper rod and a glass rod into the fire, allowing a part of each to project, the copper rod that projects out of the fire would soon become so very hot that you dare not touch it, owing to the copper conducting the heat from the fire, whereas you would be able to take hold of the projecting end of the glass rod long after the end of the glass exposed to the fire had melted. The fact is, the copper carries heat well, and the glass carries heat badly. Now with the teaching of that experiment before you, you will understand, I hope, the exact object of one or two experiments I am about to show you. Here is a piece of coarse wire gauze—I am about to place it over the flame of this Argand burner. You will notice that it lowers the flame for a moment, but almost immediately the flame dashes through the gauze (Fig. 32 A). Here is another piece of gauze, not quite so coarse as the last. I place this over the flame, and for a moment the flame cannot get through it. There, you see it is through now, but it did not pass with the same readiness that it did in the case of the other piece of gauze, which was coarser. Now, when I take a piece of fine gauze, the flame does not pass through at all until the gauze is nearly red-hot. There is plenty of gas passing all the time. If I take a still finer gauze, I shall find that the flame won't pass even when it is almost red-hot (Fig. 32 B). Plenty of gas is passing through, remember, all the time, but the flame does not pass through. Now why is it that the flame is unable to pass? The reason is this—because the metal gauze has so cooled the flame that the heat on one side is not sufficient to set fire to the gas on the other side. I must have, you see, a certain temperature to fire my gas. When therefore I experiment with a very fine piece of gauze, where I have a good deal of metal and a large conducting surface, there is no possibility of the flame passing. In fact, I have so cooled the flame by the metal gauze that it is no longer hot enough to set fire to the gas on the opposite side. I will give you one or two more illustrations of the same fact. Suppose I put upon this gauze a piece of camphor (camphor being a substance that gives off a heavy combustible vapour when heated), and then heat it, you see the camphor gas burning on the under side of the gauze, but the camphor gas on the upper side is not fired (Fig. 33). Plenty of camphor gas is being given off, but the flame of the burning camphor on the under side is not high enough to set fire to the camphor gas on the upper side, owing to the conducting power of the metal between the flame and the upper gas. There is one other experiment I should like to show you. Upon this piece of metal gauze I have piled up a small heap of gunpowder. I will place a spirit-lamp underneath the gunpowder, as you see I am now doing, and I don't suppose the gunpowder will catch fire. I see the sulphur of the gunpowder at the present moment volatilizing, but the flame, cooled by the action of the metal, is not hot enough to set fire to the gunpowder.

Fig. 33.

Fig. 34.

I showed you the steel and flint lamp—if I may call it a lamp—used by coal-miners at the time of Davy (Fig. 22). Davy set to work to invent a more satisfactory lamp than that, and the result of his experiments was the beautiful miner's lamp which I have here (Fig. 34). I regard this lamp with considerable affection, because I have been down many a coal-mine with it. This is the coal-miner's safety-lamp. The old-fashioned form of it that I have here has been much improved, but it illustrates the principle as well as, if not better than, more elaborate varieties. It is simply an oil flame covered with a gauze shade, exactly like that gauze with which I have been experimenting. I will allow a jet of coal gas to play upon this lamp, but the gas, as you see, does not catch fire. You will notice the oil flame in the lamp elongates in a curious manner. The flame of the lamp cooled by the gauze is not hot enough to set fire to the coal gas, but the appearance of the flame warns the miner, and tells him when there is danger. And that is the explanation of the beautiful miner's safety-lamp invented by Sir Humphry Davy.

Now let me once more put this fact clearly before you, that whether it is the gas flame or our farthing rushlight, whether it is our lamp or our lucifer match, if we have a flame we must have a gas to burn, and having a gas, we must heat it to, and maintain it at, a certain temperature. We have now reached a point where our tinder-box has presented us with flame. A flame is indeed the consummated work of the tinder-box.

Fig. 35.

Fig. 36.

Just let me say a few words about the grand result—the consummated work of the tinder-box. A flame is a very remarkable thing. It looks solid, but it is not solid. You will find that the inside of a flame consists of unburnt gas—gas, that is to say, not in a state of combustion at all. The only spot where true combustion takes place is the outer covering of the flame. I will try to show you some experiments illustrating this. I will take a large flame for this purpose. Here is a piece of glass tube which I have covered with ordinary white paper. Holding the covered glass tube in our large flame for a minute or two, you observe I get two rings of charred paper, corresponding to the outer envelope of the flame, whilst that portion of the paper between the black rings has not even been scorched, showing you that it is only the outer part of the flame that is burning (Fig. 35). The heat of the flame is at that part where, as I said before, the combustible gases come into contact—into collision with the atmosphere. So completely is this true, that if I take a tube, such as I have here, I can easily convey the unburnt gas in the centre of the flame away from the flame, and set fire to it, as you see, at the end of the glass tube a long distance from the flame (Fig. 36). I will place in the centre of my flame some phosphorus which is at the present moment in a state of active burning, and observe how instantly the combustion of the phosphorus ceases so soon as it gets into the centre of the flame. The crucible which contains it is cooled down immediately, and presents an entirely different appearance within the flame to what it did outside the flame. It is a curious way, perhaps you think, to stop a substance burning by putting it into a flame. Indeed I can put a heap of gunpowder inside a flame so that the outer envelope of burning gas does not ignite it (Fig. 37). There you see a heap of gunpowder in the centre of our large flame. The flame is so completely hollow that even it cannot explode the powder.

Fig. 37.

Fig. 38.

I want you, if you will, to go a step further The heat of the flame is due, as I explained in my last lecture, to the clashing of molecules. But what is the light of my candle and gas due to? The light is due to the solid matter in the flame, brought to a state of white heat or incandescence by the heat of the flame. The heat is due to the clashing of the particles, the light is due to the heated solid matter in the flame. Let me see if I can show you that. I am setting free in this bottle some hydrogen, which I am about to ignite at the end of this piece of glass tube (Fig. 38 A). I shall be a little cautious, because there is danger if my hydrogen gets mixed with air. There is my hydrogen burning; but see, it gives little or no light. But this candle flame gives light. Why? The light of the candle is due to the intensely heated solid matter in the flame; the absence of light in the hydrogen flame depends on the absence of solid matter. Let me hold clean white plates over both these flames. See the quantity of black solid matter that I am able to collect from this candle flame (Fig. 38 B). But my hydrogen yields me no soot or solid matter whatsoever (Fig. 38 A). The plate remains perfectly clean, and only a little moisture collects upon it. The light that candle gives depends upon the solid matter in the flame becoming intensely heated. If what I say be true, it follows that if I take a flame which gives no light, like this hydrogen flame (Fig. 39 A), and give it solid particles, I ought to change the non-luminous flame into a luminous one. Let us see whether this be so or not. I have here a glass tube containing a little cotton wadding (Fig. 39 B *a*), and I am about to pour on the wadding a little ether, and to make the hydrogen gas pass through the cotton wadding soaked with ether before I fire it. And now if what I have said is correct, the hydrogen flame to which I have imparted a large quantity of solid matter ought to produce a good light, and so it does! See, I have converted the flame which gave no light (Fig. 39 A) into a flame which gives an excellent light merely by incorporating solid matter with the flame (Fig. 39 B). What is more, the amount of light that a flame gives depends upon the amount or rather the number of solid particles that it contains. The more solid particles there are in the flame, the greater is the light. Let me give you an illustration of this. Here is an interesting little piece of apparatus given to my predecessor in the chair of chemistry at the London Hospital by the Augustus Harris of that day. It is one of the torches formerly used by the pantomime fairies as they descended from the realms of the carpenters. I have an alcohol flame at the top of the torch which gives me very little light. Here, you see, is an arrangement by which I can shake a quantity of solid matter (lycopodium) into the non-luminous alcohol flame. You will observe what a magnificently luminous flame I produce (Fig. 40).

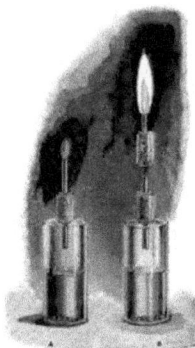

Fig. 39.

Fig. 40.

I have told you that the light of a flame is due to solid matter in the flame;[B] further, that the amount of light is due to the amount of solid matter. And now I want to show you that the kind of light is due to the kind of solid matter in the flame. Here are some pieces of cotton wadding, which I am about to saturate with alcoholic solutions of different kinds of solid matter. For instance, I have in one bottle an alcoholic solution of a lithium salt, in another of a barium, in a third of a strontium, and so on. I will set fire to all these solutions, and you see how vastly different the colours are, the colour of the flames being dependent on the various forms of solid matter that I have introduced into them.

[B] I have not forgotten Frankland's experiments on this subject, but the lectures did not admit of dealing with exceptional cases.

Thus I have shown you that the heat of our flame is due to the clashing of the two gases, and the light of the flame to the solid matter in the flame, and the kind of light to the kind of solid matter.

Well, there is another point to which I desire to refer. Light is the paint which colours bodies. You know that ordinary white light is made up of a series of beautiful colours (the spectrum), which I show you here. If I take all these spectrum or rainbow colours which are painted on this glass I can, as you see, recompose them into white light by rotating the disc with sufficient rapidity that they may get mixed together on the little screen at the back of your eye. White light then is a mixture of a number of colours.

Just ask yourselves this question. Why is this piece of ribbon white? The white light falls upon it. White light is made up of all those colours you saw just now upon the screen. The light is reflected from this ribbon exactly as it fell upon the ribbon. The whole of those colours come off together, and that ribbon is white because the whole of the colours of the spectrum are reflected at the same moment. Why is that ribbon green? The white light falls upon the ribbon—the violet, the indigo, the red, the blue, the orange, and the yellow, are absorbed by the dye of the ribbon, and you do not see them. The ribbon, as it were, drinks in all these colours, but it cannot drink in the green. And reflecting the green of the spectrum, you see that ribbon green because the ribbon is incapable of absorbing the green of the white light. Why is this ribbon red? For the same reason. It can absorb the green which the previous piece of ribbon could not absorb, but it cannot absorb the red. The fact is, colour is not an inherent property of a body. If you ask me why that ribbon is green, and why this ribbon is red, the real answer is, that the red ribbon has absorbed every colour except the red, and the green ribbon every colour except the green, not because they are of themselves red and green but because they have the power of reflecting those colours from their surfaces.

This then is the consummated work of our tinder-box. Our tinder-box set fire to the match, and the match set fire to the candle, whilst the heat and the light of the candle are the finished work of the candle that the tinder-box lighted.

The clock warns me that I must bring to an end my story of a tinder-box. To be sure, the tinder-box is a thing of the past, but I hope its story has not been altogether without teaching. Let me assure you that the failure, if failure there be, is not the fault of the story, but of the story-teller. If some day, my young friends, you desire to be great philosophers—and such desire is a high and holy ambition—be content in the first instance to listen to the familiar stories told you by the commonest of common things. There is nothing, depend upon it, too little to learn from. In time you will rise to higher efforts of thought and intellectual activity, but you will be primed for those efforts by the grasp you have secured in your studies of every-day phenomena.

"Great things are made of little things,
And little things go lessening, till at last
Comes God behind them."

THE END.

Milton Keynes UK
Ingram Content Group UK Ltd.
UKHW030839021124
450589UK00006B/672

9 789362 927965